A Layman's History

of the

STEAM RAILWAYS

OF

MALLORCA

by

Lindsay H R Fisher

I dedicate this book to all the many persons I have been privileged to call 'Friends'.

Cover picture: Tunnel entrance at Parc de la Mar, Palma. Photograph by kind permission of *Railway Magazine*.

Photographs on pages 11, 13, 14, 15, 16, 18, 19, 22 and 24 by kind permission of Lawrence G Marshall.

Photographs on pages 29 and 30 by kind permission of Snr Hausmann of Foto Balear of Carrer Constitucio, Palma, Mallorca.

Photographs on pages 45 and 48 by kind permission of Casa Bestard, Costa y Lloguera 2, Pollensa, Mallorca.

The extract of an article by Lawrence G Marshall from *Steam on the Sierra* by Peter Allen & Robert Wheeler published by Cleaver-Hulme Press Ltd, London, in 1960, is reproduced with the kind permission of the author.

© Lindsay H R Fisher 1992

ISBN 1 872356 06 0
Published by ASHBRACKEN
14 Cropwell Road   Radcliffe-on-Trent   Nottingham

Printed by Derry and Sons Limited   Nottingham
England

2

CONTENTS

The author	5
Acknowledgements	6
The island of Mallorca	7
The railways of Mallorca	9
Locomotives	15
Coaches and carriages	24
Palma station	27
Palma's underground steam railway	31
Palma to Inca line	33
The line from Consell to Alaro	34
Palma to Santanyi line	37
The Santa Maria to Felanitx line	39
Inca to Sineu, Manacor and Arta line	41
Inca to Sa Pobla line	43
The Palma to Soller railway	46
Llubi station	49
Tickets	55
Line-running during the steam era	57
A typical day on the steam railways in 1959	59
Post-steam trains on Mallorca	62
Stations on the Mallorcan and Soller railways 1991	64
Details of some Nasmyth Wilson locomotives	66
Bibliography	68

4

THE AUTHOR

Lindsay Harold Ralph Fisher was born in Portsmouth, England, on 5th March 1916. He was educated at a local secondary school and at the age of 18 he became the third (later the second) police cadet in the United Kingdom in the Plymouth City Police Force. At 21 he became a beat constable and after the Munich crisis of 1938 he transferred to the Criminal Investigation Department as records officer. In 1940 he volunteered and was accepted for flying training in the Rhodesian Air Training Group. Having to give this up for health reasons he remustered in the Royal Air Force in Rhodesia to the Provost Branch (Special Investigations). Later he was commissioned in the United Kingdom and was released from the RAF on 14th November 1946 with the rank of Flight Lieutenant.

After some eighteen months as clerk to the chief test pilot at A V Roe & Co Ltd at Lincoln, he entered the legal profession as a solicitor's clerk (later managing clerk and legal executive) with firms in Lincoln and Nottingham, specialising in litigation, crime and divorce, and he followed that career until retirement in 1980.

In 1981 he came as a widower to Mallorca where he has lived on his own ever since.

Fascinated as a pure layman by steam railways all his life, he became interested in tracing the history of steam railways in Mallorca, following his discovery of abandoned lines and stations throughout the island. After nearly five years of research in Mallorca and Europe he now presents this *Layman's History of the Steam Railways of Mallorca* which he trusts you will enjoy reading and perhaps even do some investigation of the subject yourself.

The author may be contacted as follows:

L H R Fisher, Esq
Apartments 112/3, Siesta1
Avenida Pedro Mas Reus
07408 Puerto Alcudia
Mallorca
Baleares
Spain.

ACKNOWLEDGEMENTS

My sincere thanks for very considerable assistance in the writing of this book, and the research which has gone into it, go to:

Mr & Mrs A Hallam of Calverton, Nottinghamshire, England;

Mr Peter H Wain of Keyworth, Nottinghamshire, England;

Ex-Squadron-Leader Ernie Dunsford of Welton, Lincolnshire, England;

Snr Javier Gonzalez of Pollensa, Mallorca,
    Snr Gabriel Alomar and
    Snr Juan Vicens Mestre of Alcudia, Mallorca
    for help with translations;

Snr Tomas Morrell Marques of FEVE (Ferrocarriles Españoles de Via Estracha) Palma;

Various members of the Association of Friends of the Mallorcan Railways (Asociacion Amigos del Ferrocarril de Baleares), Palma;

Peter Jennings of Swansea, South Wales;

Lawrence G Marshall of Brighton, Sussex, England for permission to use his chapter from *Steam on the Sierra;*

Warren Allen of Porthcawl, Wales for help with photocopying;

Mr J Allen of Babcock International Group plc;

Mr A J Perrin of Babcock Energy Limited;

The Local History Librarian and staff of Salford Local History Library, Peel Park, Salford, England for their assistance regarding the archives of Nasmyth Wilson.

THE ISLAND OF MALLORCA

Mallorca is the largest of the Balearic Islands which lie approximately 100 miles from the Spanish mainland in the Mediterranean Sea. The island is approximately 55 miles (88 km) long from east to west and 48 miles (80 km) from north to south. It is roughly mountainous along its west coast and has a central plain. Beaches and mountains alternate along the east coat. The island enjoys a very equable climate with summer temperatures in the high 80s and winter temperatures averaging 40 degrees.

It is not known how far back the history of the Balearic islands, and particularly Mallorca, goes. It is said that there is evidence of human existence on the islands from as far back as 3000 BC. Just south of Llucmajor at Capocorp Vell there are two stone structures (talayots) which are said to date back to pre-historic times. The origin of these structures is not definitely known for some say they are burial chambers, whilst another theory is that they were watchtowers or defensive posts.

Roman occupation

It is known, however, that in about 500 BC the Carthaginians came to Mallorca and remained until 123 BC when the Romans invaded the island and overcame them. Many names on the island have Roman origins. In fact the Roman name for Mallorca was Balearis Major, and that of Minorca was Balearis Minor. The principal Roman colonies on the island were at Palma and Alcudia. In this latter town there exist Roman ruins and a Roman theatre which are well worth seeing.

The Romans occupied the island for over 300 years and then in 426 AD, following the decline of the Roman empire, the Vandals occupied Mallorca until 902 AD when the Moors conquered the island. There still exist today many reminders of the Moorish occupation in the terracing of the land, in irrigation and in much Moorish architecture.

Under Spanish influence

In 1229 King Jamie I of Aragon landed at Santa Ponsa with an army of some 16,000 men, but despite the size of this army it took the king three months to defeat the Moors, who it is said 'defended their adopted island very strongly'.

Mallorca under Jamie I and his successors became a most

Palma in the steam railway era of the mid-1930s

important commercial and trading area with Palma a very busy Mediterranean port. During the 16th century Barbary pirates harassed the island's ports, often with bitter fighting. Because of this harassment many of the towns on the island were moved inland several miles. Hence today we find Soller and Puerto Soller; Andraitx and Puerto Andraitx; Alcudia and Puerto Alcudia; Pollensa and Puerto Pollensa.

King Jamie II of Aragon died in 1349 and upon his death Mallorca was annexed to the kingdom of Aragon. Later after the marriage of Ferdinand and Isabella in 1469, which united Aragon with Castile, Mallorca became a part of Spain as it is known today.

It is a very popular holiday island especially attractive to Britons, but more recently favoured by German, Scandinavian, French and Italian visitors and it is increasingly becoming an all-the-year-round holiday destination for mainland Spaniards.

THE RAILWAYS OF MALLORCA

No little part of Mallorca's more recent history is that of the steam railways of the island, initiated in 1868 and which continued until 12th December 1964, the last day on which a steam locomotive was operated on the railway system.

From the time of the introduction of railways on the island there were two distinct railway companies. One of these ran the main line steam railway, and the other ran the railway from Palma to Soller, which started as a steam railway but was electrified in 1929 and remains so.

The Steam Railways of Mallorca
mileages, opening and closing dates

Line	Distance km	miles	Opening date	Closing date
Palma - Inca	29	18	24 Feb 1875	*
Inca - Sineu	14	9	17 Feb 1878	1977
Sineu - Manacor	21	13	19 Apr 1879	1977
Manacor - Son Servera	20	13	30 Mar 1921	1977
Son Servera - Arta	10	6	19 Jun 1921	1977
Empalme - Sa Pobla	13	8	24 Oct 1878	1981
Santa Maria - Felanitx	42	27	7 Oct 1897	1 Jan 1968
Palma - Llucmajor	30	19	5 Oct 1916	1965
Llucmajor - Campos	13	8	28 Feb 1917	1965
Campos - Santanyi	18	12	21 Jul 1917	1965
Consell - Alaro	3	2	13 Apr 1881	24 Feb 1935
Palma - Puerto Palma	2	1	5 Feb 1931	1967
Palma - Soller	27	17	16 Apr 1912	***

Notes:
- * = Still operational.
- *** = Electrified on 14th July 1929 and still operational.
- (a) All tracks were of 3 foot gauge.
- (b) The last full day of steam on these railways was 12th December 1964.

The first lines

A decree of 1868 authorised the construction of a steam railway line from Palma, the capital, to Inca, the island's industrial town approximately in the centre of the island and housing one of Mallorca's main industries - shoe and leather goods making. The length of the authorised line was 18 miles (29 km). Following this decree the Mallorcan Railway Company was formed in 1872 and the line was completed and opened on 25th February 1875. Another company, the Mallorcan Central and South-Eastern Company, and the Mallorcan Railway Company were amalgamated in 1876 and became known as the Mallorcan Railways Company.

Subsequently the line from Inca was extended some 21 3/4 miles (35 km) to Manacor. A line 8 miles (13 km) long from Bordils Empalme, a junction created on the Inca to Manacor line, served Sa Pobla, whilst a branch line off the Palma to Inca line was established at Santa Maria and this line ran to Felanitx at a distance of 27 miles (43 km).

Another branch line

Yet another branch line of some 18 1/2 miles (30 km) from Palma to Llucmajor was opened on 4th September 1916 in the middle of the First World War in which Spain was a neutral country as it was again in the Second World War. On 31st July 1917 this branch line was extended beyond Llucmajor to Santanyi about 38 1/2 miles (62 km) from Palma.

In March 1921 the Inca to Manacor line was extended another 18 3/4 miles (30 km) to Arta. The gauge of all these steam railway lines of the island was, and is, 3 feet (91.4 cm).

It is interesting to note that at about the time steam railways were introduced to the island, the tramways of Palma operated steam locomotion. In 1917 the Mallorcan Railways purchased from Palma Tramways a Nasmyth Wilson 0-4-0 tank locomotive and a little later another 0-4-0 tank locomotive, but this one was of Orenstein & Koppel manufacture. The gauge of the tramways was 3 feet and the track consisted of flat-bottomed rail which was spiked directly to sleepers. The tramways are said to have operated until about 1936. The tram terminus was at San Antonio (in Palma) and at one time a six-times-a-day tram service ran between Palma and Arenal, the locomotion being provided latterly by a four-wheeled 12 horse-power Citroën petrol-driven car.

As you will read later, the streets of Palma also saw steam locomotives operating between the Palma main station and

Locomotive Number 4 on pilot duties at Palma
Photograph: Lawrence G Marshall
October 1957

the docks area of the city, first along the roads and later through a specially constructed tunnel under Palma.

The tunnel under Palma

In 1927 a new line was authorised to run in a tunnel 2.1 km long under the city of Palma from the city's railway station (which is still there and in use) in Plaza España to the port of Palma (or more correctly the docks area of the city). This line was only to carry goods and not passengers. At the same time the doubling of the main line from Palma to Inca was authorised. (The reader should appreciate that all the lines mentioned previously were single-line tracks.)

Both these projects and the reclamation of land near the port of Palma for the siting of the marshalling yards below Palma cathedral were begun in 1928 and completed in 1931. The tunnel entry at Palma railway station is still there to be seen, whilst the exit of the tunnel to the marshalling yards can still be found in the wall below the cathedral. However, all traces of the marshalling yards have been erased by the making of the Parc de la Mar - an artificial lake and gardens.

In certain streets in Palma the ventilation stacks of the tunnel which permitted the emission of smoke and steam from the underground railway still exist. Prior to the building of the tunnel railway under the streets of Palma all the goods imported or exported by the island were carried to or from the port of Palma by road which must have caused great congestion in the streets of Palma. Much like the present, one assumes!

Unfulfilled plans

It is interesting to note here that some fifty years ago, and persistently since then during the existence of steam railways on the island there has been mention of plans for an extension of the line from Inca to Sa Pobla to Alcudia, or even Puerto Alcudia in the north of the island a further distance of $8^{3}/_{4}$ miles (14 km) and for another extension from San Miguel on the Manacor to Arta line to the famous Cuevas del Drach (Caves of Drach) near Puerto Cristo some $4^{1}/_{2}$ miles (7 km). Comment over the years has insisted that neither of these proposed extensions would involve 'any important engineering works, both being over easy ground', but the author cannot agree with these contentions.

Closures

The dock railway in Palma was the first line to close and its use ceased as early as 1965. In 1964, the last year of full steam operations on the island, the Palma - Llucmajor - Santanyi line in the south of the island was closed down. Three years later the Santa Maria to Felanitx branch closed. The line from Empalme (Son Bordils) to Arta survived until 1977 and the shorter line from Inca to Sa Pobla did not close until 1981. The Palma to Inca line is still operating today with diesel traction, and the electrified line to Soller is also still in use.

The heyday of steam locomotion on the island was thus between about 1925 and the start of the Spanish civil war in 1936.

Locomotive Number 5 at Palma
October 1957
Photograph: Lawrence G Marshall

Locomotive Number 14 'Marratxi' at Palma sheds
October 1957   Photograph: Lawrence G Marshall

LOCOMOTIVES

England supplied the original steam locomotives and rolling stock for the steam railways of the Mallorcan Railways Company from their inauguration in 1874 to 1911, but in more recent years German and Spanish companies were the suppliers.

The slogan 'British is Best' can certainly relate to the many early steam locomotives of the Mallorcan railways which were built by Nasmyth Wilson Co Ltd of Manchester, England, and at least two were built in that company's works in Madrid. These British-built locomotives gave excellent service to the Mallorcan railways for some eighty years. Palma Station site had excellent railway workshops and amenities to service the island's steam locomotives. These workshops in 1911 built three 4-4-0 tank steam locomotives though probably the frames and boilers for these were purchased from Europe. Furthermore these workshops refurbished five of the original steam locomotives of the Mallorcan Railway Company when new frames or boilers were fitted to them.

October 1957

Locomotive Number 9 at Palma

Photograph: Lawrence G Marshall

Even after the last full day of steam trains on the island on the 12th December 1964, these stalwart steam locomotives were in daily service shunting in the sidings of Palma station. The workshops are still in existence in 1991 and are true exhibition sites, for they are just as they were when they ceased to operate and it is the author's sincere wish that the FEVE and Palma Council will very soon now preserve these workshops for posterity.

The Nasmyth Wilson locomotives

The original steam locomotives were of the 4-4-0 type with $3^1/_2$ft diameter driving wheels, 11in x 18in cylinders and in working order they weighed some $18^1/_2$ tons. All were tank engines built by Nasmyth Wilson in England.

Later the 4-4-0s had larger cylinders 13in x 18in and later still $13^1/_2$in x 19in, and in working order these weighed 27 or 28 tons. In 1876 Nasmyth Wilson built locomotives numbered 4 and 5 which were 0-6-0 tanks with 3ft 3in driving wheels and cylinders 13in x 18in, and these weighed 23 tons in working order. Eleven years later in 1887, Nasmyth Wilson built two 4-6-0 tanks with 3ft 3in diameter driving wheels and 15in x 20in cylinders weighing $35^1/_2$ tons in full working order.

Locomotive Number 19 'Lluchmayor' at Palma
October 1957   Photograph: Lawrence G Marshall

LOCO. NO. 1881/9   TEN. NO.

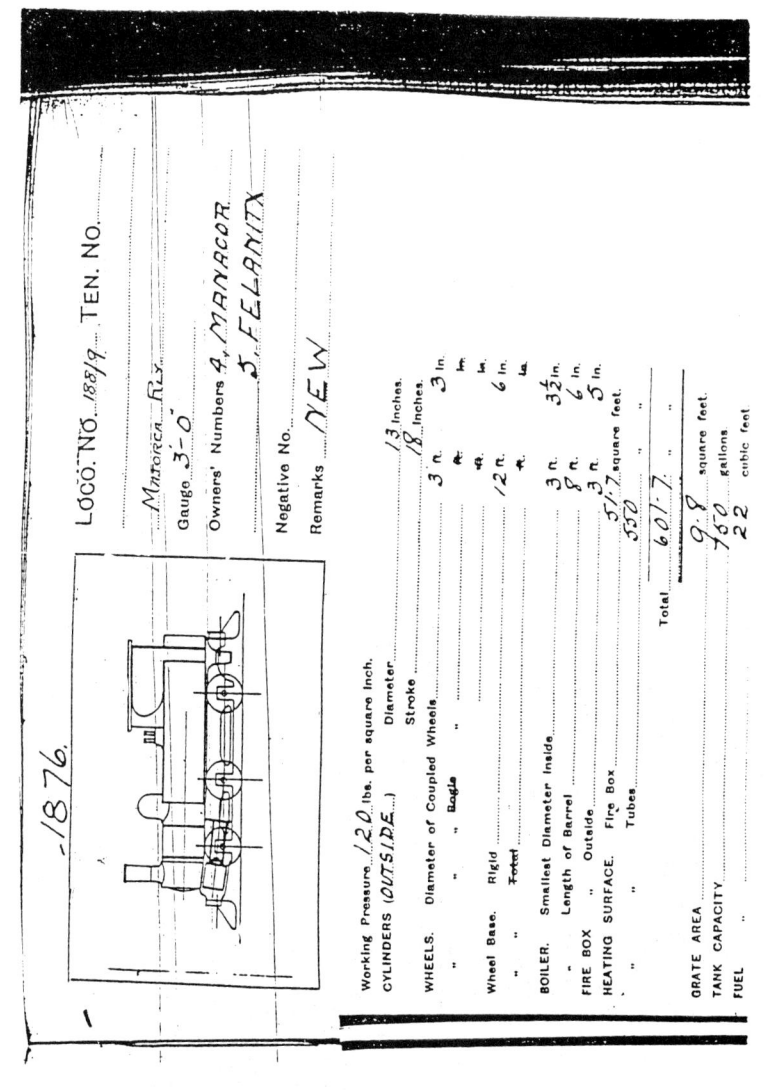

1876

Mallorca Rly.
Gauge 3'-0"
Owners' Numbers 4, MANACOR
5, FELANITX

Negative No.
Remarks NEW

Working Pressure 120 lbs. per square inch.
CYLINDERS (OUTSIDE...) Diameter.............. 13 Inches.
                        Stroke................. 18 Inches.
WHEELS. Diameter of Coupled Wheels.......... 3 ft. 3 in.
         "   " Bogie ....
Wheel Base. Rigid .... 12 ft.
             Total ....
BOILER. Smallest Diameter Inside .... 3 ft. 3½ in.
         Length of Barrel .... 8 ft. 6 in.
FIRE BOX   Outside .... 3 ft. 5 in.
HEATING SURFACE. Fire Box .... 51.7 square feet.
         "   Tubes 550
                                    Total 601.7

GRATE AREA .... 9.8 square feet.
TANK CAPACITY .... 750 gallons
FUEL .... 22 cubic feet

Nasmyth Wilson's data sheet of 1876 for locomotives for the Mallorcan Railways
Courtesy of Salford Local History Library

By 1898 some thirteen steam locomotives, all built by Nasmyth Wilson, were in service, they having supplied two further 4-6-0s of the same specifications as the 1887 ones. Again in 1911 the same company built a further two identical steam locomotives for the Mallorcan railways system.

A 0-4-0 steam locomotive was also built by Nasmyth Wilson for working traffic between Palma Town station and the port. This engine was delivered in 1889 and was fitted with 2ft 6in driving wheels and 10in x 14in cylinders, and in working order weighed 12½ tons.

Other suppliers

The next and last steam locomotives manufactured by a British company for the Mallorcan steam railways were six 2-6-2 tank engines (numbered 50 to 55) and built by Babcock & Wilcox at their works in Bilbao, Spain in 1928. They were for use for passenger traffic between Palma, Inca and Manacor, were designed to haul 100 tons at an average speed of 30 to 40 kilometres per hour, and they carried sufficient water for the single journey. They were fitted with electric headlights, feed water heaters and purifiers. They had coupled wheels 3ft 7in in diameter, cylinders 14½in x 21¾in, and in working order they weighed 46 tons.

Locomotive Number 27 'Arta' at Palma

October 1957

Photograph: Lawrence G Marshall

Dimensions of some Mallorcan Steam Locomotives

Date	Wheel pattern	Diameter of driving wheels (ft)	(in)	Cylinders Diameter (in)	Stroke (in)	Weight in working order (ton)
1874	4-4-0T	3	6	11	18	18.5
1876	0-6-0T	3	3	13	18	23
1877	4-4-0T	3	6	13	18	27
1881	4-4-0T	3	6	13.5	19	28
1887	4-6-0T	3	3	15	20	35.5
1889	0-4-0T	3	6	10	14	12.5
1891	4-6-0T	3	3	15	20	35.5
1898	4-4-0T	3	6	13.5	19	not known
1911	4-6-0T	3	3	15	20	35.5
1917	2-6-0T	3	5	14.25	19.75	34

Further details of some of the Nasmyth Wilson locomotives appear on pages 66 and 67.

Locomotive Number 16 'Porreras' at Palma sheds
October 1957
Photograph: Lawrence G Marshall

Summary of steam locomotives used by the Mallorcan Railways Company

Date	Builder	Type	Number	Name	Retired'
1874	Nasmyth Wilson	4-4-0T	1	Majorca	
1874	Nasmyth Wilson	4-4-0T	2	Palma	
1874	Nasmyth Wilson	4-4-0T	3	Inca	
1876	Nasmyth Wilson	0-6-0T	4	Manacor	1960
1876	Nasmyth Wilson	0-6-0T	5	Felanitix	1960
1877	Nasmyth Wilson	4-4-0T	6	Sineu	Rebuilt 1911
1877	Nasmyth Wilson	4-4-0T	7	La Puebla (Sa Pobla)	Rebuilt 1911
1877	Nasmyth Wilson	4-4-0T	8	Santa Maria	Rebuilt 1911
1877	Nasmyth Wilson	4-4-0T	9	Benisalem	Rebuilt 1911
1881	Nasmyth Wilson	4-4-0T	10	Muro	Rebuilt 1911
1881	Nasmyth Wilson	4-4-0T	11	Petra (Renumbered as No.18 in 1929)	1929
1887	Nasmyth Wilson	4-6-0T	12	St Juan	
1887	Nasmyth Wilson	4-6-0T	13	Lloseta	
1889	Nasmyth Wilson	0-4-0T	6	Acquired from Palma Electric Tramways c.1917	
1891	Nasmyth Wilson	4-6-0T	14	Marratxi	1960(?)
1891	Nasmyth Wilson	4-6-0T	15	Alaro	
1898	Nasmyth Wilson	4-4-0T	16	Porreras	1960
1898	Nasmyth Wilson	4-4-0T	17	Montuiri	
1902	Palma Works	4-6-0T?	18	Salinas (Renumbered as No.22; c.1911.)	1960(?)
1903	Palma Works	4-4-0T	19	España (Renumbered as No.23)	
1911	Nasmyth Wilson	4-6-0T	20	Algaida (Renumbered as No.18 in 1917; renumbered as No.11 in 1929)	1960(?)
1911	Nasmyth Wilson	4-6-0T	21	Santa Eugenia (Renumbered as No.10 c.1917)	1917(?)
?		4-4-0T	2		
1911[R]	Palma Works	4-4-0T	24	Coll	1960
1911[R]	Palma Works	4-4-0T	25	San Miguel	
1911[R]	Palma Works	4-4-0T	26	San Lorenzo	
1911[R]	Palma Works	4-4-0T	27	Arta	1960(?)
1911[R]	Palma Works	4-4-0T	28	San Servera	
1911	La Maquinista, Barcelona	2-6-0T	7	Acquired in 1944 from Soller Railway	1960
1911	La Maquinista, Barcelona	2-6-0T	8	Acquired in 1944 from Soller Railway	1960
1911	La Maquinista, Barcelona	2-6-0T	9	Acquired in 1944 from Soller Railway	1960
?	Orenstein & Koppel	0-4-0T	7	Acquired from Palma Electric Tramways c.1917	1944

20

Summary of steam locomotives used by the Mallorcan Railways Company

Date	Builder	Type	Number	Name	Retired[1]
1917	La Maquinista, Barcelona	0-4-0T	6		1960(?)
1917	La Maquinista, Barcelona	2-6-0T	19	Lluchmayor	1960(?)
1917	La Maquinista, Barcelona	2-6-0T	20	Campos	1960(?)
1917	La Maquinista, Barcelona	2-6-0T	21	Santanyi	
1926	Krupp, Germany	2-6-0T	30		
1926	Krupp, Germany	2-6-0T	31		
1926	Krupp, Germany	2-6-0T	32		
1926	Krupp, Germany	2-6-0T	33		
1926	Krupp, Germany	2-6-0T	34		
1926	Krupp, Germany	2-6-0T	35		1960(?)
1930	Babcock & Wilcox (Bilbao)	2-6-0T	50		1960(?)
1930	Babcock & Wilcox (Bilbao)	2-6-0T	51		1960(?)
1930	Babcock & Wilcox (Bilbao)	2-6-0T	52		1960(?)
1930	Babcock & Wilcox (Bilbao)	2-6-0T	53		1960(?)
1930	Babcock & Wilcox (Bilbao)	2-6-0T	54		1960(?)
1930	Babcock & Wilcox (Bilbao)	2-6-0T	55		1960(?)

Steam Locomotives of the Soller Railway Company

Date	Builder	Type	Number	
1891	Falcon Engine Co	0-4-0T		Tram
1911	La Maquinista, Barcelona	2-6-0T	1	No.7 Mallorcan Railways from 1944
1911	La Maquinista, Barcelona	2-6-0T	2	No.8 Mallorcan Railways from 1944
1911	La Maquinista, Barcelona	2-6-0T	3	No.9 Mallorcan Railways from 1944
1911	La Maquinista, Barcelona	2-6-0T	4	

Notes on the Table.
The (R) after the date indicates an extensive rebuild of an earlier locomotive.

Purchases from Palma Tramways

In 1917 the Mallorcan Railways Company purchased from the Palma Tramways Company a Nasmyth Wilson-built small four-coupled steam engine. Some time later another 0-4-0 steam locomotive, this one built by Orenstein & Koppel, was also purchased from the same source. Both these steam locomotives had been used by the Palma Tramways over their street trackage between Palma Town station and the docks, and such trackage was used by the Mallorcan Railways Company until the underground railway from Palma Town station to the docks was inaugurated. It is known that the Orenstein & Koppel locomotive was in constant use until the mid 1960s, but the fate of the Nasmyth Wilson locomotive is not known.

Memories of 1921

I have seen it recorded by a visitor to Mallorca in 1921, when remember the population of the whole island was less than a quarter of a million people and the population of Palma was only about 70,000, that the steam trains on the island's system usually had First and Second Class coaches, though some had Third Class accommodation comprising what were virtually cattle trucks with a bar down the middle

Locomotive Number 55 at Inca heading the 14.25 from Palma to Arta
October 1957   Photograph: Lawrence G Marshall

to which passengers merely held on. Trains, the visitor recorded, were rarely on time, possibly due to 'fearful overloading' and he states that on one occasion he travelled on a train consisting of 21 carriages pulled by one 2-6-0 tank locomotive. This same visitor asserts that when he visited the island in 1921 there was an 'Express Train' which ran non-stop from Palma to Inca, a track distance of 29km, in 30 minutes and he adds that he recorded a speed of 65 kilometres per hour on this train.

Permanent way

In the 1930s the permanent way (the track) was built of 30 kilogram per metre (60 pounds per yard) flat-bottomed rails generally 10 metres (32ft 10in) long. These were spiked to the wooden sleepers which were stone ballasted. Train speeds rarely exceeded 40 miles per hour (50 kilometres per hour).

On the double track between Palma and Santa Maria (then the only track doubled on the system) trains ran on the right hand track, following the pattern of vehicular traffic on the island which travels on the right hand side of the road.

On the whole railway system on the island, the steepest gradient was 1 in 67 while curvature was no more than approximately 800 feet radius except through station loops.

In the mid-1930s, at the height of the era of steam railways on the island, trains generally left Palma station in the morning with corresponding return trains in the afternoon and evening, and most trains comprised a loco-motive and four-wheeled carriages, or carriages and wagons as demand dictated.

The average 'booked speed' on the 58^1/$_2$ miles run from Palma to Arta, including stops, was 23^1/$_2$ miles per hour. The fastest start-to-stop run between Palma and Santa Maria (15km or 9^1/$_4$ miles) was 20 minutes, this being made in both directions despite the fact that there is a substantial rise from Palma to Santa Maria.

Trains were often 'double-headed' out of Palma as far as Santa Maria or Inca.

Most of the early steam locomotives on the island were re-boilered at the company's workshops at Palma station, many of the locomotives doing Trojan service for eighty years and more and ending their working life on the shunting service at Palma Town station.

All steam trains on the system carried names as well as numbers - the names being those of towns served by the system. (See pages 64 and 65.)

COACHES AND CARRIAGES

The first coaches and carriages on the steam railways of Mallorca were made by Brown Marshalls & Co Ltd of Birmingham, England, and they were of the British-type four-wheeler compartment coaches plus a few saloons. A full length footboard at wheel-axle level ran the whole length of the coach and a false roof above the main roof provided protection from the sun. Four-wheel brake vans combined with the coaches. Most of the rolling stock lasted 70 to 80 years, proving the solidity of the designs.

All rolling stock at the beginning of the system was fitted with a centre buffer with a semi-automatic coupling, but from the 1920s the couplings were supplemented on later stock by a screw coupling under the buffer. Later still both locomotive and passenger coaches were fitted with vacuum brakes.

It is significant that the locomotives were only fitted with a connection at the rear which indicates that they were turned at the end of each journey. Most terminal stations and lines still in existence show how the turn-round was achieved.

Three carriages and a brake van at Palma
October 1957　　　　　　　　　　Photograph: Lawrence G Marshall

However, I have only found the sites of two turntables on the whole system. The only complete surviving turntable is at Palma station, whilst at what is left of Santanyi station there is a pit, circular in shape, and reinforced round its rim, which I am sure was a turntable when the station was operational.

New luxury coaches

In 1930 fourteen bogie corridor coaches built by Material Movil y Construccione of Saragossa were purchased for the system. The design of these coaches was quite remarkable when one considers the problems involved in putting a side-corridor into a three-foot gauge coach. First and Second Class accommodation was provided and the standard of comfort in these new Second Class side-corridor coaches meant that a small supplement was charged on the fare.

Rail motor cars and coaches

In 1936 the Mallorcan Railways Company owned four rail motor cars which at that time were used for operating some of the less-heavily loaded services and for special services. The Railways Company also then owned some 567 goods wagons and some 85 passenger coaches, most of the latter being four-wheeled coaches which had been operational ever since the steam railways first ran on the island. Also in 1936 the Railways Company had fourteen modern bogie corridor coaches used on the main line services. At that time the coach stock was either varnished teak or painted brown.

In the late 1930s First and Second Class accommodation was provided on all lines on the island except the Palma, Santa Maria and Felanitx line which was Second Class only. If one travelled in the Second Class on a train composed of modern coaches a supplement was payable as accommodation was far superior to other Second Class accommodation offered.

All passenger and freight stock used was fitted with centre buffers with automatic hook couplers as well as side chains, passenger stock also having screw couplers and vacuum braking.

In 1936 the Mallorcan Railways Company also owned a number of motor buses and operated a road service to outlying towns and villages in conjunction with train arrivals at the various stations.

Railcars were introduced into the system in 1926 and these were built by Berliet Co Auxiliar FCC Besain and Construcione Naval Sestrao, Bilbao and Essaugen Co, Eskalduna.

It is interesting to note that the company of Babcock & Wilcox in Bilbao, Spain, was established originally by Babcock & Wilcox Ltd of the United Kingdom and that the Bilbao works built some 530 steam locomotives between 1923 and 1961 for the steam railways of Spain.

Livery

The Nasmyth Wilson steam locomotives used on the system were painted green and were originally lined-out in black. In 1936 all steam locomotives on the island were painted green and most carried names of towns or villages served by the railways. By the end of the era of steam locomotion green was still in use but black was apparently the predominant colour.

Steam locomotives used on the system but which were built on the continent of Europe, in Germany or Spain, appear to have been always painted in unlined black.

When the Spanish government took over the running of the steam railways of Mallorca in 1951 the livery was changed for coaches and carriages to green below the waist and pale grey above whereas previously these were painted brown.

Reasons for decline

In my view several matters contributed to the decline in the prosperity of the Mallorcan railways in the 1930s, viz, the Spanish civil war, the increase in road traffic and a general reduction in world trade. The war in Spain itself caused little trouble to the island but commerce was badly affected. The long-term effect of these influences was that on 1st August 1951 the management of the system was taken over by Explotacion de Ferrocarriles por el Estado (the Railway Department of the Ministry of Works).

In 1930 it is recorded that the Mallorcan railways (excluding the Soller Railway) possessed 32 locomotives, 610 trucks, 64 passenger coaches and 35 wagons. Another interesting statistic from that year is that the railways had some 870 agents on the island.

It is sad to record that in 1990 a tragic fire, believed to have been caused by vagrants, destroyed a large shed immediately adjacent to Palma station where several of the original coaches of the Mallorcan railways were stored. These coaches were also completely destroyed.

PALMA STATION

Palma station, which has been in existence ever since there were railways on the island of Mallorca, lies at the junction of Plaza España and Carrer Marques de Fuensanta. As one enters the station area from Plaza España, on the right of the terminal track ending in buffers, is a solid-built building which houses the administrative offices of FEVE (Ferro carriles Españoles de Via Estracha) the national railway company of Spain which has since 1959 been responsible for the railways of Mallorca, with the exception of the Soller Railway, both formerly in steam and today in diesel power. On this building, just below the roof, is a building-block bearing the Roman numerals MDCCCLXV (1865) - the date of the building - and a similar block appears on the solid-built single-storey building on the left which houses the booking office and staff rooms. Straight ahead of one runs the double track out of Palma disappearing into the distance in a perfectly straight line. Ahead of one, but away to the left, lie

Palma station: a view of the station platform, ticket office and staff rooms

the old engine sheds with a turntable in front of them and further left the railway workshops described elsewhere in this book. Immediately behind the booking office on the left was, until 1990, the large carriage shed which until that date contained quite a few of the older carriages of the Mallorcan Railways, which in that year were destroyed in the fire referred to on page 26.

It is interesting to note that on the front of the building to the right housing the administrative offices is a commemorative plaque which reads:

C/a de Los Ferrocarriles
de Mallorca
Caides por Disos y por La Patria
Barcello Tomas (Gabriel)
Pascual Pomar (Miguel)
Sureda Payeras (Mateo)
1936   Presentes!   1939

to the memory of three members of the staff of the Mallorcan Railways who lost their lives in the Spanish civil war of 1936 to 1939.

The 'Hostel Terminus', (which is a residential hostel or hotel), lying between Palma station proper and the Soller railway station but a hundred yards apart, is of exactly the same construction as the buildings in Palma station and it carries a building block showing that it was built in the same year as the station buildings.

Points operation

All points on the rails of the station are individually and manually operated, while all signalling is by lights a metre or so above the tracks and these signals are mechanically operated.

An aerial view of Palma station in 1946 looking towards Plaza España

Photograph courtesy of Snr Hausmann of Foto Balear, Carrer Constitucio, Palma, Mallorca

An aerial view of Palma harbour in 1946. The entrance to the underground railway is to the right and below Palma cathedral. The track layout serving the dock area can just be discerned.
Photograph courtesy of Snr Hausmann of Foto Balear, Carrer Constitucio, Palma, Mallorca

PALMA'S UNDERGROUND STEAM RAILWAY

Prior to 1927 steam trains ran through the streets of Palma from the main station in Plaza España to the port, and often they consisted of five trucks or wagons. The line was a pure freight line, either to take exports from the island to the mainland by way of the ships using the port of Palma, or to carry all sorts of freight from the ships in the port to the main Palma station for distribution throughout the island.

In 1927 a tunnel line with a single track was authorised from Palma main station to the port area and its marshalling yards. The entrance to the tunnel can be seen under an accumulation of rubbish at Palma main station along the wall of the station on Carrer Marques de Fuentsanta, while the emergence of the tunnel-line at the port area is still visible on the cliff face just below the cathedral. What is now the Parc de la Mar was the site of the marshalling yards for this railway at the port of Palma.

Palma station: the entrance to the long-since abandoned underground railway from the station to the port area

Air vents which allowed steam and smoke to rise out of the tunnel can still be seen on certain streets in Palma and consist of a metal box approximately 80cm by 40cm with a cover on top supported at each corner and about 20cm from the top of the box. This tunnel line was inaugurated in 1931 and ceased to operate in 1965. It was approximately 2 kilometres in length and like its predecessor it was a pure freight line.

If one visits Palma today it is not difficult to appreciate what traffic chaos (even in those days) must have been created by having most of the docks' freight traffic carried by a steam locomotive, trucks and wagons along a 3-foot gauge line before the tunnel was built.

Palma trams

The Palma Tramway Company was formed in 1914 and the first trams were drawn by mules until 1916 when the system was converted to electrical power. In 1950 the Palma city network of tramways comprised lines serving the Can Capas, Santa Catalina, Porto Pi, Son Roca, Cenova, La Soledad, Cloliseo and Establiments districts and it was eventually expanded to reach Cas Catala and El Arenal. It is recorded that by 1959 the company carried some 4,512,341 passengers. In February 1958 motor buses arrived in Palma and within a few months most of the tramlines were removed from the roads and sold as scrap.

Santa Maria station on the Palma to Inca line which is still in use.
The line to Felanitx branched off eastwards here

PALMA TO INCA LINE

This line, which throughout this book is referred to as the main line, was opened in 1875. It covers a distance of some 28.6km and is still in service as it was originally laid down. However, I must argue against various references to this line contained in other authors' books and articles which declare that this line has been a double line since 1931.

I came to live permanently in Mallorca in 1981 and then there was no double track from Santa Maria to Inca, nor, if my memory serves me right, for some distance from Santa Maria towards Palma. In fact I saw the line for some distance south of Santa Maria and from Santa Maria to Inca built into a double line, for I was greatly fascinated by a very modern tracklaying machine which I understand was brought from France to lay the second track.

The stations on this line are at Palma, Pont d'Inca, Marratxi, Santa Maria, Alaro & Consell, Benisalem, Lloseta and Inca. All have very substantial buildings in stone and brick. The stations at Palma, Pont d'Inca, Santa Maria, Benisalem, Lloseta and Inca are all in built-up areas of these various towns, but Marratxi is a short distance out of the town proper, whilst Alaro & Consell (one station which serves both towns) is approximately equidistant between these two towns and 1km from each. All stations of course have a station master.

There are many small roads over which this line travels and since I have lived in Mallorca most of them have been fitted with electrically controlled booms to close the road just prior to trains approaching. Previously they were 'manned' by a local resident who placed a chain across the road whenever a train was approaching.

An interesting point is that there were no signals of the drop-arm type, common throughout Europe, on any of the railway lines in Mallorca during the age of steam.

THE LINE FROM CONSELL TO ALARO
The steam railway that never was

The main line from Palma to Inca became operational in 1875 and the inhabitants of the small town of Alaro, some 3.4 km north-west of Consell, one of the stations on the main line, had hoped that the main line would have gone through Alaro, but this was not to be. Alaro, besides its inhabitants, had a coal mine and several factories nearby which would have used any railway between Consell and Alaro.

An attempt was made to float a company to construct such a railway link, but this never got off the ground. However, the various persons interested in creating such a rail link were not deterred. A single line track was constructed from the main line station at Consell and since a steam locomotive could not be afforded a tram-type conveyance carriage to transport people and goods was purchased, as were two mules to pull it. Passengers sat on seats on the roof of the carriage and the line started operation on 22nd May 1881.

Over the following thirty or so years various ideas to improve the line were aired but never acted upon until in 1921 the mules were 'pensioned off' and two motor tractors were bought to pull the coach to and from Alaro and the main line station at Consell, now renamed Alaro & Consell. This

Alaro & Consell station on the Palma to Inca line which is still in use. This is the only 'double-barrel' named station in the whole of Mallorca

tractor-hauled railway existed until 12th March 1941.

Between 1944 and 1945 a set of rails approximately 1200 metres in length was laid from the main line station near Consell to the coal, or more accurately lignite mine near Alaro and steam locomotives of Ferrocarriles de Mallorca took the coal wagons to and from the mine and the main line station. This coal-only line persisted until 1951.

The timetables of this railway over the years are most interesting to study.

In 1891 the timetable was:

Alaro to Consell	Consell to Alaro
5.57 (a)	6.20 (c)
8.03 (b)	8.22 (d)
8.57 (a)	9.16 (c)
14.33 (b)	14.57 (d)
17.15 (b)	17.37 (d)
19.13 (a)	19.31 (c)

(a) With a connection to Palma. (b) With a connection to Inca.
(c) With a connection with a train coming from Inca.
(d) With a connection with a train coming from Palma.

In 1931 it appears that this line was what is described as a tramway and its running-times were aligned to trains stopping at Consell station on the main Palma to Manacor line, which of course ran through Inca.

In 1934 the timetable was:

Alaro to Consell	Consell to Alaro
5.35 (a)	
7.35 (b)	
7.55 (c)	8.15 (d)
8.55 (b)	9.15 (e)
13.00 (a)	
14.15 (a)	14.35 (d)
15.00 (b)	15.25 (e)
18.55 (c)	19.40 (f)

(a) With a connection to Palma. (b) With a connection to Inca.
(c) With a connection to trains to Inca and to Palma.
(d) With a connection with a train from Inca.
(e) With a connection with a train from Palma.
(f) With a connection with trains from both Inca and Palma.

I surveyed the line thoroughly in early 1991 to find that all traces of the tracks had been removed long ago. Any goods-sheds and stables from the time mules were used to draw the carriages, and the later garages for the motor tractors, at the site of 'Alaro & Consell' station had long since been demolished, whilst the site of the station at Alaro was now a substantially built Spanish home.

Santanyi station: the terminus of the Palma to Santanyi line.
On the right are the remains of a double engine shed. In the foreground to the right and below ground level are the remains of a turntable. To the left of the other building, a storage shed, is the site of the station building which has since been demolished

PALMA TO SANTANYI LINE

This branch line opened in 1916 virtually along the south coast of the island and was an extension from Palma towards the south-east corner of the island. Originally the line was constructed from Palma to Llucmajor, a distance of 30.8 kilometres, with stations at Coll d'en Rebassa, Arenal and of course at Llucmajor. I carried out a survey of this branch line in early 1991 and I found that the station at Coll d'en Rebassa still exists. It is well preserved and now belongs to the local education authority

The stations which did exist at Arenal and Llucmajor have long since been demolished and the sites built over with modern concrete buildings.

The branch line was extended to Santanyi, a distance of 62 kilometres from Palma, and the extension was opened on 1st July 1917 and operated until 4th March 1964. There were stations on this extended branch line at Campos, Banos de San Juan, Ses Salines and at Santanyi itself. My survey of this extension of the branch line revealed that the very well preserved station at Campos is now the home of a local Spanish family. The Banos de San Juan, which are spa baths, are in fact some little distance from the station of that name.

Campos station on the Palma to Santanyi line.
The building is now a very comfortable home. This is a view from the line side and the station name is hidden behind the white awning.

The station there now bears no identification by name, but it is very obviously a well-built stone building identical to many still existing throughout the island. The station at Ses Salines is still solid to outside observation but vandalised inside. The terminus station at Santanyi has long since been demolished, but at the station site there still exists a two-bay engine shed, unfortunately fast-deteriorating in condition. There is also a single-storey storage building in fair condition and a circular pit amidst the grass. This pit is about $2 \frac{1}{2}$ feet deep with a reinforced edge and it appears that it originally housed a turntable.

Timetables during the era of steam on the island show that there were halts at San Francisco (11 kilometres from Palma), Las Cadenas (12 km), El Palmer (48 km) and Llomparts (58 km). Stops were also made at Canteras (51 km) and Banos (52km). Other more important stations were situated at Coll d'en Rebassa, Arenal, Llucmajor, Ses Salines and Campos.

After the branch line closed in 1964 the tracks were removed and as with all the lines on the island the steam locomotives were shipped to the mainland and cut up for scrap.

Santa Eugenia station on the Santa Maria to Felanitx line.
The building is now a recreational centre

THE SANTA MARIA TO FELANITX LINE

In 1897 a branch line off the main Palma to Inca line was constructed from Santa Maria, which lies approximately half way between those two towns, to Felanitx at a distance of some 42.8 kilometres. The new branch line ran west to east across the centre of Mallorca and had stations at Santa Eugenia, Algaida, Montuiri, Porreras and a terminal at Felanitx.

This branch-line closed in 1967 and thereafter the tracks were removed and unfortunately the stations in several instances were left to vandals. However all stations still exist today as follows:

Santa Eugenia station has been used for some time as a council leisure centre. The track obviously ran in front of the building for the name of the station is over the main door onto the road upon which one approaches the station, whilst at the rear of the station a swimming pool has been built. The line crosses the road between Santa Eugenia and Algaida at which point is a well-preserved building which was formerly an unnamed halt.

The station at **Algaida** is now a private house, extensions having been constructed onto the original station building, particularly on what was the track side of the station.

The entrance to what was Felanitx station.
The building is now a medical centre

Montuiri station is also well-preserved as a private house with orange trees now growing where the track was.

The station at **Porreras** presents rather a different picture having been very badly vandalised.

However the terminal station at **Felanitx**, which is the only single-storey station building on the branch line, has been put to excellent use and extended for it is now a medical centre, Estacion Enologica. Regrettably the other station buildings comprising three storage sheds and a double engine shed are in a very sad state of repair, the latter being used as a stable for horses or mules in early 1991.

As far as this branch line is concerned it must be said that the stations were all built either on the immediate outskirts of the towns or in the towns themselves which the line served.

Manacor station on the Inca to Arta line seen from the line side.
Note the station name over the central door

INCA TO SINEU, MANACOR AND ARTA LINE

This branch line was inaugurated in 1878 and originally ran only as far as Sineu some 13.9 kilometres from Inca. An extension of 21.3 kilometres was built in 1879 to Manacor, but it was not until 1921 that the line was extended from Manacor to Arta a further distance of some 30.7 kilometres. This line was eventually abandoned in 1977.

On the original branch line there was one station between Inca and Sineu, namely Empalme, otherwise known as San Bordils, which later became a junction when a further branch line was extended northwards to Sa Pobla. This extension is dealt with later as a separate branch line.

When the branch line was extended towards Manacor a station was built at Petra and later still, when the line was further extended from Manacor to Arta, stations were built on this extension at San Lorenzo, San Miguel and Son Servera, and of course at the terminus at Arta itself. The line was last used by steam locomotives in 1964, but all the stations on this whole branch line have preservation orders on them and so are protected as far as possible from vandals, and all the tracks of the line are still in situ.

As early as 1936 proposals were voiced to extend the line even further from San Miguel south-eastwards to the Caves of Drach, or Cuevas del Drach, and Porto Cristo, but such a project never materialised.

Timetables during the era of steam on the island show that trains on this line stopped at San Juan, 49 kilometres from Palma, between Petra and Sineu, though no station is traceable today. Between 1964 and 1977 diesel traction was used on this line.

On a survey of this branch line in 1990 I found the various stations in a variety of conditions as follows:

Inca station is of course still in full use, being now the terminus of the Palma to Inca line, the only surviving line from the heyday of steam locomotion on the island.

Empalme is still as it was, complete with all tracks as when it was abandoned by steam in 1964.

The station at **Sineu**, a substantial building of two and a half storeys on the main road in the town, with the rail track on the other side of the buildings, is now a bar and café.

Petra station, comprising a two storey building, stands abandoned and slightly vandalised.

Manacor station, which is a central building of two

storeys with single-storey wings on either side of the main station building, is, subject to some slight vandalism, as it was when the branch line was abandoned by steam in 1964.

San Lorenzo and **San Miguel** stations are identical buildings consisting of one storey with a goods storage shed close by and are well-preserved.

Son Servera station is of the type just described.

Arta station, on the eastern side of the town, is a large two-storey building with end windows above the second floor and with various outbuildings, again well-preserved.

Son Servera station on the Inca to Arta line viewed from what was the track side. Note the station name over the central doorway

INCA TO SA POBLA LINE

This line became operational in 1878, the line running approximately eastwards from Inca to Empalme and then approximately north to Sa Pobla through Llubi and Muro, a distance of some 13.1 kilometres.

Empalme was a junction with no true station as such, but with various railway buildings. At least one of these has the name 'Empalme' in a building-block of the main entrance. There are several switch points on the track which is still in situ. It is understood that preservation orders are in existence relating to this branch, its track and stations.

The station at **Llubi** is some considerable distance of about 2 km from the town itself so users of the line had a longish walk home if they were passengers and a longish drive if they were farmers or traders using the line for freight carriage.

Empalme station: this railway building stands where the line to Sa Pobla turns northwards from the Inca, Manacor and Arta line. Note that in 1991 the tracks were still in position

Llubi is easily approached by road off the Inca to Sineu road. The station is a single-storey building, somewhat vandalised, but one can easily picture in one's mind activity at the station during the era of steam on the Mallorcan railways. On the opposite side of the track to the station, and with access from a country road which runs at right angles to the track just outside the station, is what was a two-storey goods building, again extensively vandalised. This station is dealt with more fully later (page 49).

Where the track crosses the road immediately outside the station there still stands the usual warning sign to road traffic that trains cross the road ahead. Like all other stations throughout the steam railway system on the island Llubi has its own well.

Muro station is also some distance from the town. It is a well preserved two-storey building with doorways and windows well boarded up and thus it is not vandalised like so many other stations on the system. There is the usual goods storage building on the same side of the track as the station.

The terminus of this branch line at **Sa Pobla** is enclosed in a stout brick wall at a street junction on the north-east corner of the town. A yard surrounds the single-storey station building which is well preserved with doorways and windows blocked up. As the track approaches the station there is, on the town side of the track, the only elevated water tank I have found on the island. It is a tank approximately 10 feet by 6 feet by 4 feet and it stands some 17 feet from ground level on top of a very substantially-built stone block base. The track is in situ as at the date of closure of this branch line in 1981, but remember the last steam train travelled this line in 1964.

I surveyed and photographed this terminal station at Sa Pobla in 1989 when all the tracks at the station and those stretching back towards Muro were visible, but on revisiting the station a few weeks later I found a considerable length of track had been submerged under asphalt to provide a newly-opened supermarket with a car park.

My research of this line showed that as long ago as 1936 an extension of this line to Alcudia or Puerto Alcudia was suggested and over the intervening years this idea has surfaced several times but has never got off the drawing board and now in my opinion never will.

An electric coach, built circa 1929, on the Soller railway
Photograph courtesy of Casa Bestard, Costa y Lloguera 2, Pollensa, Mallorca

THE PALMA TO SOLLER RAILWAY

This railway runs from Palma, using a small station almost adjacent to the main line station, through Son Sardina and Bunyola to Soller. Then there is an electric tramway from Soller station to Puerto Soller.

The line which is, and always has been privately owned by Ferrocariles de Palma Soller SA, was planned and approved in 1906 and in the following year work began on the longest tunnel on the line. This is Tunel Major cutting through the northern mountain range, the Sierra Norte, and which has a length of 2857 metres. There are thirteen tunnels altogether on the line covering a length of approximately 5 kilometres out of a total length of 27 kilometres.

The line opened on 16th April 1912 and it is interesting to note that the government of the day offered financial assistance to minor railways such as this one provided that the construction covered more than 30 kilometres. The line from Palma to Soller was only 27.1 kilometres long, so to obtain the government financial assistance the rail company constructed the still existing and operational tramway from Soller station to Puerto Soller. The tramway is some 4.9 kilometres in length and thus they built a railway and tramway of some 32 kilometres which obtained for the company the government financial assistance it sought.

The line is a masterpiece of railway engineering. It rises at a point 20 kilometres from Palma, above the town of Soller, to a summit of 323 metres (1060 feet) above sea level. The Palma terminal is 43 metres (141 feet) above sea level whilst that at Soller is 41 metres (134 feet) above sea level. On the journey between the termini one enjoys a steady climb through wonderful mountainous country and, as already mentioned, through several tunnels. The line rises high above the town of Soller and thence, on a very winding track, drops nearly 1000 feet to Soller station through fine views of sea and mountains. Between 1912 and 1929, when the line was electrified, the railway was worked by 2-6-0 tank steam locomotives. Nos. 1 to 4 of these locomotives were built by La Majuinista Terreste y Maritima of Barcelona in 1911. Nos. 1, 2 and 3 were transferred to the Mallorcan Railways in 1944. An 0-4-0 tram was built in 1891 by Falcon Engineering Co and purchased by the line company when the tramway was built and ran from outside Soller station to Puerto Soller.

The opening of this line caused the abandonment of the

stagecoach service between Palma and Soller. This service is referred to in George Sand's novel *Winter in Majorca* in which she describes her stay with her two children and Frederick Chopin in Valdemossa between November 1838 and April 1839.

At Palma the Soller Railway Company had only a car shed but the Soller terminus housed the line's shed and works. Each steam locomotive on this line weighed 32 tons in working order and burned a mixture of wood and coal dust. They were capable of a rare turn of speed and it is said to be recorded that the 6 kilometre run from Son Sardina to Palma took just six minutes.

The carriages are, for most of them still survive in service, eight-wheeled bogie types and the trains consisted of first and second class coaches. The former were divided into three compartments and contained 32 small armchairs throughout the coach and there was a balcony at each end of the coach. The third class coaches had no compartments but had seats for 64 passengers. At the end of each coach there was a central buffer, whilst screw couplings were standard and Westinghouse brakes were applied to all wheels. Trucks were eight or four-wheeled and were fitted with a hand-brake in addition to screw couplings and Westinghouse braking. In addition they were fitted with an 'adapter' so that they could be used on the steam rails of Mallorca. Trucks on this line were always put on to the end of the train for there were no freight trains as such.

In the era of steam on this line the journey from Palma to Soller lasted approximately 65 minutes. The first stop out of Palma at Son Sardina, a distance of some 6 kilometres, was reached in ten minutes. The other interim stop on this line at Bunyola, approximately 15 kilometres from Palma, was reached in 30 minutes. The longest tunnel on the line takes between six and seven minutes to traverse.

Various documents and writings, which I have seen during my research of this book, when concluding their write-up of this railway line nearly always say: 'This line appears to have always been run in a most professional way and it is known for its punctuality of service.' My own experiences certainly fully endorse these statements.

The Soller Railway is said to be the only privately owned public railway in Spain and it is run by Ferrocarril de Soller SA of Soller which is an entirely different company from the Ferrocarriles Espanoles de Via Estecha (FEVE), which today runs the other railway in Mallorca.

An electric tram on the tramway between Soller and Puerto Soller.
The tram ride lasts some twelve to fifteen minutes
Photograph courtesy of Casa Bestard, Costa y Lloguera 2, Pollensa, Mallorca

LLUBI STATION

This station on the Inca to Sa Pobla line is situated very close to a road which crosses the line at right angles just near the Sa Pobla end of the station. Whilst researching this book I visited Llubi station to photograph the buildings and I was able to enter a building on the far side of the track to the station which was apparently a goods storage building in the heyday of steam railways in Mallorca and which was badly vandalised. Amongst the debris of the demolished interior walls of the building I discovered various railway documents, some badly mouse or rat-eaten, but some of which I have been able to repair and these are still in my possession.

These documents shed a great deal of light on the state of the line commercially in 1945 and also in 1964. One document - *Clasificacion de Mercancias y Ganados Transportados de dicha procedencia duren Enero of 1964 (Classification of Merchandise and Livestock carried during January 1964)* shows that goods carried that month were as follows:

Goods	Charge (pesetas)
Alcoholic drinks	3.40
40 kg fruit (fresh or dry)	6.80
600 kg miscellaneous goods	22.70
1 pig	13.20
Total income	46.10

This makes the total income from all goods transported to or from Llubi station for the whole month of January 1964 to be less than 50 pesetas. I wonder whether this would have paid the station master's salary for that particular month.

A second document appears to show the amount of total cash transactions, presumably from ticket sales and goods charges for the month of January 1964 at 6327.30 pesetas. This sum is split into three ten-day periods of the month, details of which are as follows:

1st to 10th January 1964	2440.10	pesetas
11th to 20th January 1964	1802.55	pesetas
21st to 31st January 1964	2084.65	pesetas

Llubi station on the Inca to Sa Pobla line.
Note that in this general view taken in 1991 the tracks are still visible

Yet another document covers the period of 11th to 20th January 1968, which was after the steam trains had ceased running on the island, and this shows that thirteen general, or ordinary, tickets were issued. The return to the company is shown as 19.91 pesetas plus insurances of 2.50 pesetas. The insurances were compulsorily charged and collected by the railway company but these were not income to the company. Thus the total receipts were just 22.41 pesetas. But 'No!' - all fares must be what an Englishman would call 'rounded up to the nearest peseta' and thus the total transaction is 23.00 pesetas.

Another document, a very large one, gives a very detailed return of all tickets issued at Llubi station and this is for the whole month of April 1964 when the line was still under steam. The issue of tickets is again shown using the ten day method as follows:

1st to 10th April 1964	265
11th to 20th April 1964	141
21st to 31st April 1964	164
Total	570

The cash value of these tickets is shown as a total of 6078.00 pesetas, whilst the goods returns for the whole month of April 1964 show that income from that source was 6077.65 pesetas, a very similar figure.

Another large document shows the issue of tickets to other specified stations on the system during April 1945 as below:

Llubi to Empalme	1st of the month	7	
	5th of the month	6	
	10th of the month	3	Total = 16
Llubi to Sineu	11th to 20th of the month	1	
Llubi to S. Juan	21st to 30th of the month	2	

This same document also gives details of the issue of tickets between Llubi and Inca. The types of tickets are described as first-class, second-class for adults, and separately children, military authorisations etc. The figures of the values of tickets, in pesetas, issued for travel from Llubi to Inca are as follows:

	1st to 10th April 1945	11th to 20th April 1945	21st to 30th April 1945
Adults first-class	4.95	3.20	---
second-class	128.70	78.10	59.40
Authorisations	0.90	0.90	2.10
Others	0.66	0.66	---
Totals	135.21	82.86	61.50

Total for the month of April 1945 was 279.57 pesetas.

A final document in my possession is a typed list of fares per kilometre and, for example, the break-down of a fare of 17 pesetas for a 63 kilometre journey and for a fare of 22 pesetas charged for a journey of 80 kilometres is shown as:

	60 kilometre journey	80 kilometre journey
The Railway Company's Charge (or 'true' fare)	15.75	20.00
plus various tax and insurance totalling	2.20	1.40
'Rounding up'	0.05	0.60
making a total fare of	17.00	22.00

Ferrocarriles de Mallorca   Estación de _____

CLASIFICACION de mercancías y ganados transportados de dicha procedencia durante _____ de 19___

NUMERO Y DESIGNACION DE LOS GRUPOS	PESO Kilógramos	PRODUCTO Pesetas Cts.
1 Aceites industriales, resinas, betunes y grasas		
2 Aceites		
3 Alcoholes	10	0.40
4 Algarrobas		
5 Almendras y almendrón		
6 Azúcar		
7 Carnes frescas y saladas		
8 Calzados confeccionados		
9 Cemento, cal y yeso		
10 Carburo de calcio		
11 Drogas y productos químicos		
12 Estiércoles		
13 Forrajes y piensos para el ganado y remolacha		
14 Frutas frescas y secas	40	6.50
15 Hortalizas		
16 Huevos		
17 Lana en bruto, sucia o en churre		
18 Lana lavada, peinada o cardada e hilada para tejidos		
19 Leche		
20 Manteca y embutidos		
21 Minerales en general		
22 Obra de palma y espartería		
23 Orujos		
24 Papel de todas clases		
25 Pescado, sus salazones y conservas		
26 Pulpas y conservas de todas clases		
27 Pieles curtidas y sin curtir		
28 Sal		
29 Tejidos de todas clases		
30 Tierras y sillares		
31 Vinos		
32 Volatería al peso		
33 Demás mercancías no expresadas E	600	22.70
SUMA Y SIGUE	650	32.90

Front of *Clasificacion de Mercancias y Ganados Transportados de dicha procedencia duren Enero of 1964 (Classification of Merchandise and Livestock carried during January 1964)*. The original in the possession of the author measures 15.8cm by 21.7cm (6¼in by 8½in)

52

NUMERO Y DESIGNACION DE LOS GRUPOS	PESO Kilógramos	PRODUCTO Pesetas Cts.
SUMA ANTERIOR	550	32 70
34 Abonos (1), E .		
35 Arroz, E .		
36 Carbones Minerales, E .		
37 » vegetales, E .		
38 Cereales de todas clases (2) E		
39 Harinas, E		
40 Leños, E .		
41 Legumbres (3), E .		
42 Envases vacios de todas clases, E.		
43 Maderas para el ramo de construcción, E .		
44 Tablillas desarmadas para cajas de madera, E .		
45 *Tubérculos y raices, E* .		
TOTALES	650	33 70

(1) No se continuarán en este grupo las materias que sirven para elaboración de los abonos artificiales como la Fosforita, Pirita, etc etc. por tener que figurar en el grupo de Minerales.
(2) Se entenderá por cereales de todas clases, el Alforfón, Alpiste, Avena, Cebada, Centeno, Escauda, Escaña, Maiz, Mijo, Panizo, Sorga, Trigo y Zaina.
(3) Por legumbres las que sirven para la alimentación humana de uso general y frecuente.

GANADOS

CLASES DE GANADO	Núm. de Cabezas	PRODUCTO Pesetas Cts.
Vacuno .		
Caballar, Mular y Asnal		
Lanar y Cabrio		
De Cerda	1	13 20
TOTALES	1	13 20

, 31 de Enero de 1964

El Jefe de la Estación,

Back of *Clasificacion de Mercancias y Ganados Transportados de dicha procedencia duren Enero of 1964 (Classification of Merchandise and Livestock carried during January 1964)*. The original in the possession of the author measures 15.8cm by 21.7cm (6¼in by 8½in)

Before closing this chapter I think the reader may well be very interested to know more details of the work carried out by all station masters beyond attending to at least three train arrivals and the same number of departures seven days a week.

Amongst the railway documents I found at Llubi station were the following:

1. Forms 28.5cm by 18.0cm (11^{1}/$_{4}$in by 6^{2}/$_{10}$in) covering a thirty or thirty-one day month with returns each day in twelve columns.

2. A book of forms measuring 30.5cm by 21.5cm (12in by 8^{1}/$_{2}$in) requiring returns every 10 or 11 days, each day with some 21 columns on each page. The period covered in the book I possess was from the second 10-day period in August of one year to the first 10-day period in April of the following year. This was a return made in duplicate so that the pages in the book that I have are carbon copies of the original returns.

3. Monthly return forms measuring 75cm wide by 25.5cm (29^{1}/$_{2}$in by 10in) printed on one side of the paper only and with returns made daily. The forms comprise two sets of columns, one above the other dividing the paper into two horizontally and going right across the width of the paper. The upper half of the form has 39 columns and two marginal entries, while the lower half has 34 columns. Every one of the 73 columns had to be totalled.

4. A monthly return measuring 16cm by 22cm (6^{1}/$_{2}$in by 8^{1}/$_{2}$in) which is the form illustrated in this chapter. As the reader can see this comprises some 33 items on the front of the form with each item detailed and those applicable requiring an entry as to the weight and another as to the cartage charge. Then on the reverse side of the form there were a further 12 items calling for the same details. In the lower half of the page there were four spaces for entries of livestock being carried detailing the type of livestock, their weight and the charge for carriage.

5. A further large form for monthly returns and measuring 60.5cm by 29.5cm (24in by 11^{1}/$_{2}$in) with three 10 or 11-day periods one above the other and each day's entry containing 31 columns.

Hence it would appear that in the era of steam on the island a station master, in addition to his other duties, had to be a very efficient book-keeper!

TICKETS

Today at most railway stations throughout the world one can purchase a great variety of tickets for one's journey e.g. single, return, monthly, season, senior citizen, etc, etc.

My research on this book in Mallorca threw up a very interesting style of ticket used by the Mallorcan Railways during the era of steam. From 1943 to be exact. The ticket is called *Billete Kilometrico*, which translates literally as *Kilometres ticket*. The reader will note from the illustrations below and on page 56 that the ticket is dated 1943 and expires on the 1st day of March 1944, and permits thirteen 5 kilometre journeys first-class. The principal condition of the ticket is that the holder had to present the ticket one half hour before the departure of the train upon which he wished to travel.

Presumably the holder of the ticket presented it at the ticket office at his station of departure, said that he wished to travel on a certain train due shortly to another certain station on the system, say 15 kilometres distant, and the

FERROCARRILES DE MALLORCA
BILLETE KILOMETRICO

Los adjuntos cupones deberán canjearse por billetes complementarios en las taquillas de la Compañía, antes de emprender viaje durante la media hora anterior a la salida de los trenes en que deba utilizarse y no se admitirá ningún cupón que no esté adherido a esta hoja.

ACCIONES Nº 13807

AÑO 1943 Caduca en 1.º de Marzo de 1944	9 1.ª clase 5 Kilómetros	4 1.ª clase 5 Kilómetros
13 1.ª clase 5 Kilómetros	8 1.ª clase 5 Kilómetros	3 1.ª clase 5 Kilómetros
12 1.ª clase 5 Kilómetros	7 1.ª clase 5 Kilómetros	2 1.ª clase 5 Kilómetros
11 1.ª clase 5 Kilómetros	6 1.ª clase 5 Kilómetros	1 1.ª clase 5 Kilómetros
10 1.ª clase 5 Kilómetros	5 1.ª clase 5 Kilómetros	1.ª clase AÑO 1943 Caduca en 1.º de Marzo de 1944

The front of a *Billete Kilometrico* now in the possession of the author. The original on brown paper measures 15.3cm by 11cm (6in by 4¼in)

ticket clerk or more likely the station master would hand the holder a ticket for his particular journey and cancel three of the 5 kilometre permits printed on the ticket as illustrated.

Equally fascinating I find is the Table of Kilometre Distances printed on the reverse side of the 13-journey ticket, illustrated below. This table relates to every one of the stations on the steam railway system on Mallorca in 1943

I feel sure that the reader will agree with me that the organisation and initiative shown by the issue of this 13-journey ticket nearly fifty years ago is something to be applauded.

NB The station called 'La Puebla' is called throughout this book 'Sa Pobla', the two names being alternatives.

The back of a *Billete Kilometrico* now in the possession of the author. The original on brown paper measures 15.3cm by 11cm (6in by 4$^{1}/_{4}$in)

LINE-RUNNING OF THE MALLORCAN RAILWAYS DURING THE STEAM ERA

Pre-1926 the rail journey from Palma to Arta took just over two hours with stops to take on water at Inca and Manacor. As the steam locomotives aged, the stretch of line between Manacor and Arta became an increasingly difficult climb through the mountains calling for more sturdy and powerful 2-6-0 tank locomotives and these were purchased from Krupps of Germany in 1926.

The Palma, Inca, Sa Pobla and Arta lines

The timetables of the island's railways for about 1960, at a time when steam locomotives were still a common sight on the system, show that nine trains each weekday ran from Palma to Inca and the journey time was about 45 minutes. Three of those trains ran through Inca to Sa Pobla, the total journey time from Palma to Sa Pobla being 1 hour 15 minutes. Four more trains ran from Inca via Manacor, to which town the journey time from Palma was 1 hour 25 minutes, and then on to the terminus at Arta, the total journey time from Palma being two hours.

Two further trains ran on Sundays and 'festivos' days, or national holidays, only from Palma to Inca in the evening. One of these trains went on through to Sa Pobla and the other went through to Manacor only. Three of the weekday trains on 'festivos' days ran one hour later than usual. The four daily return runs from Arta on weekdays left at 7.55, 13.12 and 17.50, the evening train on 'festivos' and holidays leaving at 18.50.

A weekday train left Manacor, originating from there, for Palma via Inca at 6.00 a.m. On weekdays trains left Arta for Palma via Manacor and Inca at 7.55, 13.12 and 17.50, the latter at 18.50 on 'festivos' days and Sundays. Weekday trains left Sa Pobla at 7.00, 12.30 and 17.33.

One of the Palma to Arta trains was a 'split shift' turn based at Manacor including a trip to Palma, returning to Arta, back again to Palma and with a final run to Manacor. A typical train in 1957, for instance, on the Palma to Arta run consisted of a 2-6-2 tank locomotive, brake van, three bogie-coaches, four 4-wheeled coaches, five or six vans and a final brake van. Quite a train!

Palma to Felanitx

Two trains a day ran on this line of a weekday, one in early afternoon and the other in mid-evening and the journey time was 1 hour 30 minutes. The evening train ran one hour later on 'festivos' days between 1st October and the following 31st May, and two hours later from 1st June to 30th September. Return journeys from Felanitx to Palma were at 7.30 and 17.30.

Palma and Santanyi

On weekdays three trains a day ran on this line, the journey time being 1 hour and 20 minutes. An additional mid-evening train also ran on 'festivos' days and Sundays only and this evening train ran an hour later than during the rest of the year on holidays between June and September, showing great consideration for the wishes of passengers.

It is interesting to note that a normal train on this line in 1957 comprised usually a railcar and a postal carriage, in addition to passenger carriages and wagons.

A TYPICAL DAY ON THE STEAM RAILWAYS OF MALLORCA IN 1957

I have always believed that contemporary notes of any event are always the truest description of the event being witnessed and in the course of my research into this book I came across an account of a visit to Mallorca by a very well known railway researcher and author, Mr Lawrence G Marshall of Brighton, Sussex, England, who visited the island in 1957 and shortly afterwards wrote an account of what he had seen. This is reproduced below with his kind permission:

"Where to Señor?"
"Mallorcan Railway station, please."
"But Señor!", said the driver of my vast American taxi as we roared away from the hotel, "tourists always use the coaches for sightseeing, the trains are very slow and very dirty. Only soldiers and the poor Spaniards use them."
"Trains are very slow and very dirty" - that must mean steam trains, I said to myself with a sigh of satisfaction. Soon we swung off the cobbled streets of Palma into a cross between a chicken-yard and an open air market, and there in the background I saw the magic, if faded, words Estacion de los Ferrocarriles de Mallorca *[Station of the Railways of Mallorca].*
Fighting my way through two huge queues of people patiently waiting for tickets for the two morning trains to distant Arta and Sa Pobla I finally arrived at the platform. On the left hand side of the departure platform stood two diminutive blue and cream railcars each with a wonderfully antiquated brown four-wheeled coach attached to take the inevitable overflow. These were the two morning departures to Santanyi and Felanitx, two branches now given over entirely to railcars for passenger working. The main departure platform was completely occupied with a ten-coach train of these four-wheeled coaches followed by four vans and several open wagons. At the head of this delightful array stood a diminutive 2-6-0 tank locomotive endeavouring to raise the energy to move this cavalcade and obliterating the area in smoke at the same time - small wonder most of the Mallorcan steam engines are dirty, for they only burn coal dust! Like so many narrow gauge lines, the Mallorcan Railways are run by the state, which means that it is essential to keep the

line running, but as little money as possible must be spent in doing so! The train for Arta was a far more presumptuous affair and consisted of six modern bogie corridor coaches hauled by one of the six Babcock & Wilcox 2-6-2 tanks, also in a woebegone and paintless condition.

Once these four trains had departed the railway's two little Nasmyth Wilson 0-6-0 tanks with stovepipe chimneys, Nos.4 and 5, began the most vigorous shunting campaign I have ever seen; coaches, vans and wagons being treated with the utmost disdain - small wonder that the little coaches showed more bare wood than paint on their ceilings. Old Nos.4 and 5, dating from 1876, are among the oldest British-built narrow-gauge locomotives still in use in Spain; these together with No.27, which appeared to be the last survivor of the once ubiquitous 4-4-0 tanks - two of which were built on the island - seemed to share the Palma station pilot and shunting duties between them.

Time now for a look round the two locomotive roundhouses; the first and largest seemed to house all the active locomotives, and in steam were three of the large modern Babcock & Wilcox 2-6-2 tanks, two of the class of five Krupps 2-6-0 tanks delivered in 1926 and one of the four surviving Nasmyth Wilson 4-6-0 tanks. These latter engines are spending the remainder of their active lives on troop and freight workings, two of them dating back to 1887. The only other engine in steam was No.9, one of four top-heavy looking 2-6-0 tanks built by the Spanish firm of La Maquinista Terrestre y Maritima between 1911 and 1917. These engines were some of the very first to be built by the firm and as the heavy Sa Pobla train had shown earlier on, they are still very capable little machines. No.9 incidentally, was purchased from the neighbouring Soller Railway when that line was electrified in 1929. The smaller roundhouse adjoining the workshops seemed to house all less active members of each class, together with the little railcars. During 1956 two French-built railcars were delivered to the island, and these have proved the ideal answer for the less heavily loaded Felanitx and Santanyi trains. Scattered all round the area were the decaying remains of various locomotives, mainly 4-4-0 tanks, but including the sixth member of the Krupps 2-6-0 tanks, No.32, and the tiny Orenstein & Koppel 0-4-0 tank, No.7, built as recently as 1921 for the harbour branch.

Leaving Palma the main line runs due north; at the far

end of the yards the harbour branch makes a trailing connection from the right, and shortly after the Santanyi branch, eighteen miles long, curves away to the right. The main line, doubled in 1931, now bears to the north-east and passes through the most delightful scenery - vineyards, fig trees, giant prickly-pear cactus and fields of bright red pimentos abound, whilst the line itself is bordered by great lines of almond trees. To enhance the picture still further, the landscape for the first few miles out of Palma is dotted with gaily painted little windmills, providing the necessary irrigation for the crops. At Santa Maria the twenty-seven miles long Felanitx branch turns away from us to the east and from here onwards the soil gradually becomes more barren and less fertile.

After Inca, the ancient capital of the island, the line becomes single and more tortuous, and at Empalme the Sa Pobla section swings off north while we continue eastwards across the centre of Mallorca to Manacor, famous for the subterranean Dragon Caves and fine imitation pearls. After Manacor the line turns north again and we finally arrive after almost four hours' journey at Arta, fifty-nine miles from Palma.

Fares are very cheap but, unless one travels first-class, severe overcrowding must be endured. The main lines to Arta and Sa Pobla have two trains a day each way, and the Felanitx and Santanyi branches three railcars. On the two main lines all trains are mixed, and on the two branches separate freight trains are run as required. Unfortunately none of the four lines quite reaches the sea, but, nevertheless, the Mallorcan Railways provide an admirable means of seeing the beautiful island; long may they continue to do so.

Unfortunately Mr Marshall's wish has not been fulfilled for there were various closures as follows:

Palma to Arta (beyond Inca)	closed in 1977
Palma to Felanitx	closed in 1967
Palma to Santanyi	closed in 1964
Inca to Sa Pobla	closed in 1981

The only line still operational of all those described by Mr Marshall is the main line from Palma to Inca.

POST-STEAM TRAINS ON MALLORCA

It is interesting to note that as early as 1926 the Mallorcan Railways were using a 40 horse-power four-wheeled Berliet railcar. The railcar unfortunately caught fire in 1936 and was totally destroyed.

In 1930 three similar railcars but with bodywork by Companie Auxiliar FCC de Beasin and with a 40 horse-power De Dion engine were purchased, and these railcars worked well into the 1960s.

In 1956 four railcars built by Easslingen of Germany and later three built by Eskalduna were purchased. They looked very modern and were capable of 75 kilometres per hour and of multiple operation.

The power cars had a small baggage area but the trailers were for passengers only. There were trailers with a postal compartment and that fact was displayed by a yellow livery at the appropriate end of the trailer.

In the Spanish section of the *Thomas Cook Continental Timetable for 1986* there are entries covering the two remaining railways of Mallorca. Details of these appear below and on the following page.

Table 437. PALMA - INCA
Narrow gauge Majorca Rly (FEVE) One class only
Distance 29 km, journey 36 minutes.

From Palma and Inca:

6.00PO	7.00	8.00	8.40 ⊗
9.00Z	9.20⊗	10.00	11.00
12.00	12.40⊗	13.00Z	13.20⊗
14.00	14.40⊗	15.00Z	15.20⊗
16.00	17.00	18.00	19.00
20.00	20.40⊗	20.40⊗	21.00Z
21.20⊗	22.00PX		

Z - Sats. and Suns.
⊗ - Monday to Friday only, except holidays.
PO - from Palma only.
PX - from Inca only.

Table 437 - contd. PALMA - SOLLER
Narrow gauge (1,2 class) Soller Rly.
Distance 28 km, journey 55 minutes.

Departures from Palma	Departures from Soller
Weekdays	Weekdays
8.00	6.45
10.40 *	9.15
13.00	11.50
15.15	14.10
19.45	18.20
Sundays	Sundays
8.00	6.45
10.40 *	9.15
13.00	11.50
15.15	14.10
19.45	18.20
	21.00

Connecting service by tram from Soller to Puerto Soller

* a special 'tourist train' which stops at Mirador del Pujol d'en Banya, this being a special stop to enable enthusiasts to take some stunning photographs of the line and the town of Soller.

In March 1992 the running times were as shown above, and the fare from Palma to Soller was

1st Class single	465 pesetas
2nd Class single	330 pesetas
Return	660 pesetas

STATIONS OF THE MALLORCAN AND SOLLER RAILWAYS IN 1991

Name	Line	Open	Notes
Alcudia	IP		Extension never built
Algaida	PF		Used as private house
Alaro	AC		
Arenal	PS		Demolished
Arta	IA		Terminus. Still standing
Banos de San Juan	PS		Still standing
Bunyola	SR	**	
Benisalem	PI	**	
Campos	PS		Used as private house
Canteras	PS		Halt
Coll d'en Rebassa	PS		Used for local education
Consell	PI	**	
Consell	AC		
Cuevas del Drach	IA		Extension never built
El Palmer	PS		Halt
Empalme	IA		Still standing. Tracks in situ
Empalme	IP		Still standing. Tracks in situ
Felanitx	PF		Terminus. Now a medical centre
Inca	PI	**	
Las Cadenas	PS		Halt
Llomparts	PS		Halt
Lloseta	PI	**	
Llubi	IP		Still standing
Llucmajor	PS		Demolished
Manacor	IA		Still standing
Marratxi	PI	**	
Montuiri	PF		Used as private house
Muro	IP		Well preserved
Palma	PI	**	
Palma	SR	**	
Palma (Docks)	PD		Demolished
Petra	IA		Still standing (vandalised)
Pont d'Inca	PI	**	
Porreras	PF		Still standing (vandalised)
Puerto Soller	ST	**	
San Francisco	PS		Halt.
San Juan	IA		Still standing
San Lorenzo	IA		Still standing
San Miguel	IA		Still standing
Santa Eugenia	PF		Used as leisure centre
Santa Maria	PI	**	

64

STATIONS OF THE MALLORCAN AND SOLLER RAILWAYS
IN 1991 (continued)

Name	Line Open	Notes
Santanyi	PS	Terminus. Demolished
Sa Pobla	IP	Terminus. Still standing
Ses Salines	PS	Still standing (vandalised)
Sineu	IA	Used as bar and café
Soller	ST **	
Soller	SR **	
Son Sardina	SR **	
Son Servera	IA	Still standing

Lines:
- AC = Alaro & Consell Railway
- IA = Inca to Arta
- IP = Inca to Sa Pobla
- PI = Palma to Inca
- PF = Palma to Felanitx
- PS = Palma to Santanyi
- SR = Soller Railway
- ST = Soller tramway

**   Station in use in 1991.

Station names in **bold** type are names carried by steam locomotives on the Mallorcan and Soller Railways.

DETAILS OF SOME NASMYTH WILSON LOCOMOTIVES BUILT FOR THE MALLORCAN RAILWAYS

		1874	1876	1881	1887
Date					
Wheel arrangement		4-4-0T	0-6-0T	4-4-0T	4-6-0T
Loco. No.		174/6	188/9	205/6	320/1
Gauge		3'-0"	3'-0"	3'-0"	3'-0"
Owners' Numbers		1 Mallorca	4 Manacor	10 Muro	12 San Juan
		2 Palma	5 Felanitx	11 Petra	13 Lloseta
		3 Inca			
Negative No.		174	----	Use 174	
Remarks		New	New	Similar to 194-7	New
Working Pressure	lbs. per square inch	120	120	130	130
Cylinders		Outside	Outside	Outside	Outside
Diameter	inches	11	13	13½	15
Stroke	inches	18	18	19	20
Wheels					
Diameter of Coupled Wheels	ft. in.	3'-6"	3'-3"	3'-6"	3'-3"
Diameter of Bogie Wheels	ft. in.	2'-0"	---	2'-0"	2'-0"
Wheel Base Rigid	ft. in.	6'-9"	12'-6"	7'-9"	12'-1"
Total	ft. in.	14'-10"	---	16'-4½"	19'-5½"
Boiler Smallest Diameter Inside	ft. in.	3'-0¼"	3'-3½"	3'-5½"	3'-7"
Length of Barrel	ft. in.	8'-0"	8'-6"	8'-9"	10'-3"
Fire Box Length Outside	ft. in.	2'-11"	3'-5"	3'-9"	4'-4"
Heating Surface Fire Box	square feet	41	51.7	62	75
Tubes	square feet	391.7	550	628	787
Total	square feet	432.7	601.7	628	862
Grate Area	square feet	8	9.8	10.8	12.7
Tank capacity	gallons	600	750	850	1000
Fuel	cubic feet	18	22	23.5	

DETAILS OF SOME NASMYTH WILSON LOCOMOTIVES BUILT FOR THE MALLORCAN RAILWAYS (continued)

		1891 4-6-0T	1898 4-4-0T	1911 4-6-0T
Date				
Wheel arrangement				
Loco. No.		414/5	526/7	934-4
Gauge		3'-0"	3'-0"	3'-0"
Owners' Numbers		14 Marratxí	16 Porreras	20 Algaida
		15 Alaró	17 Montuiri	21 Santa Eugenia
Negative No.		----	Use 174	943
Remarks		Same as 205/6	Similar to 194-7	Same as 414
Working Pressure	lbs. per square inch	130	140	130
Cylinders		Outside	Outside	Outside
Diameter	inches	15	13½	15
Stroke	inches	20	19	20
Wheels				
Diameter of Coupled Wheels	ft. in.	3'-3"	3'-6"	3'-3"
Diameter of Bogie Wheels	ft. in.	2'-0"	2'-0"	2'-0"
Wheel Base Rigid	ft. in.	12'-1"	7'-9"	12'-1"
Total	ft. in.	19'-5½"	16'-4½"	19'-5½"
Boiler Smallest Diameter Inside	ft. in.	3'-7"	3'-5½"	3'-7"
Length of Barrel	ft. in.	10'-3"	8'-9"	10'-3"
Fire Box Length Outside	ft. in.	4'-4"	3'-9"	4'-4"
Heating Surface Fire Box	square feet	75	62	75
Tubes	square feet	787	628	787
Total	square feet	862	628	862
Grate Area	square feet	12.7	10.8	12.7
Tank capacity	gallons	1000	850	1000
Fuel	cubic feet	33		

BIBLIOGRAPHY

Articles entitled *The Railways of Mallorca (Balearic Isles)* by R D Unwin of Barcelona, published in Vol IX; No 54 (25th May 1921) and Vol IX; No 56 (25th June 1921) of *The Locomotive News and Railway Notes*.

Extract relating to the railways of Majorca from *Mediterranean Island Railways* by P M Kalla-Bishop, published in 1970 by David & Charles.

Pocket Book Es. *Minor Railways and Tramways in Eastern Spain* by J Morley & K P Plant, published by The Birmingham Locomotive Club (Industrial Locomotive Information Section) in 1963.

Extract of an article by Lawrence G Marshall from *Steam on the Sierra* by Peter Allen & Robert Wheeler published by Cleaver-Hulme Press Ltd, London in 1960.

A series of articles over five issues of *The Continental Modeller* between July 1986 and March 1987 entitled *Ferrocarril de Mallorca* by Giles Barnabe.

An article entitled *Mallorca and its Railway System* (from notes by the late Major S A Forbes) contained on pages 142 to 144 of *The Locomotive* dated 15th August 1942.

An article entitled *Mallorca and its Railways* by W A Willox in *The Railway Magazine* for April 1936.

El ferrocarril a Mallorca La iarda Mallorquina by Nicholau S Canellas, issued by the Govern Balear, Conselleria de Treball y Transports (a limited edition) 1990.